Cosmic Energy
What Everybody Needs to Know

Other publications by the same Author

Cosmic Energy and the nature's way in Health and Medicine" by Partridge Publishing, 2015

Cosmic Energy
What Everybody Needs to Know

The universal energy for Human life

Ko Paandu

LitPrime Solutions
East Brunswick Office Evolution
1 Tower Center Boulevard, Ste 1510
East Brunswick, NJ 08816
www.litprime.com
Phone: 1-800-981-9893

© 2025 Ko Paandu. All rights reserved.

No part of this book may be reproduced, stored in a retrieval system, or transmitted by any means without the written permission of the author.

Published by LitPrime Solutions: 03/11/2025

ISBN: 979-8-88703-436-2(sc)
ISBN: 979-8-88703-437-9(e)

Library of Congress Control Number: 2024920706

Any people depicted in stock imagery provided by iStock are models, and such images are being used for illustrative purposes only.

Certain stock imagery © iStock.

Because of the dynamic nature of the Internet, any web addresses or links contained in this book may have changed since publication and may no longer be valid. The views expressed in this work are solely those of the author and do not necessarily reflect the views of the publisher, and the publisher hereby disclaims any responsibility for them.

Contents

Author's note vii
Acknowledgment ix
Preface xi

1. Introduction 1
2. COSMOS 4
3. A Day in One's Life 7
4. Cosmic Energy 16
5. Cosmic rays- History 23
6. Life Around us 28
7. Wind around us 35
8. Positive and Negative ions 43
9. Wind and Health 47
10. Cosmic Energy in Health 51
11. Cosmic Energy in Medicine 54

Appendix-1 (Physics of Cosmic Energy) 57
References 69

Author's note

Many essential subjects and events in our life go unobserved, even though they are used by us. Hence for many, this subject may be new and also surprising. For ease of reading I have kept the technical explanation to a minimum. Most of the sources I referred to in the text can be found in the References.

Acknowledgment

My thanks go to my friend Dr. Pratul Kumar Chatterjea, Environmental protection engineer and Safety analyst. An unexpected chance to go into the field of negative ions through the book 'Ion effect' by Fred Soyka. Cover page briefs the full book in one sentence 'Revolutionary discoveries reveal that electrically charged particles in the air may control your Moods, Health and Sense of well being'

Timely reviews on the write-up and several new ideas suggested by my friends Mr. Ma.Elangovan, Mr. V.Adithya and Mr. M.Rajan. Thanks to R.Krishnamoorthy, whose review helped me to improve certain vital points. I owe my thanks to Ms. P.Thenral, for her creative ideas in organizing the subjects and in its lucid presentation with illustrations.

I owe my thanks to Mr. Boopathyrajasekar for editing and forming the contents to this format.

To my parents Kolandai Ponna Kavunder and Theetthiammaal, parents, play their unknown and innumerable part for various reasons.

To my Wife Everest for her patience during my work on this book.

Preface

The purpose of this book, "Cosmic Energy: What Everybody Needs to Know" is to make the public aware of the basic life energy around us. We have not much realized about this hidden energy. Nature has power, energy, and remedy for all the living things on earth. We should have awareness of those subjects to utilize them to the best of our ability. Human race need not create anything new once he learns how to use what is available in nature. We get all what is required to live when we respect and conserve nature.

A few chapters from the book "Cosmic Energy and the nature's way in Health and Medicine" by the same author, published by Partridge Publishing, India (2015), have been taken and updated.

An interest in cosmic energy came my way in a minor incident a few years after treatment for my health problem. I was tryed Modern medicine for a year, which could not resolve. In an unexpected circumstance, I met a medical practitioner who used to prescribe alternative medicines for many health problems. On taking that medicine, improvement happened within ten days, and subsequently, there was a good improvement in six months. This experience made me interested in knowing what it contains, how it works, and how it is better than other medicines that could not solve my problems.

My knowledge in the field of medicine was relatively poor, and my interest in that subject faded over the years once my health problem

PREFACE

had subsided. When I came across the book 'Ion Effect,' the effect of ions in the air over humans re-kindled an interest in me.

That made me understand ions linked with the medicine I chewed a decade ago. When I started reading the book, it appeared as if it mirrored my experiences from my childhood.

The book 'Ion effect' explains the author's experiences in Geneva. The author, a Canadian businessman, suffered from health problems. After his ailments in Geneva, he found after a lot of research that his issues were due to a lack of negative ions in the atmosphere. I could relate many experiences of the author to a particular type of environment in Geneva and my experience to a similar environment during a certain period of the year in my village.

I could realize the effect energy in life. Many hidden energies play a major role without our knowledge. They have different effects than we see in materials. One needs to know a little about the effect of these energies in our life.

One of the hidden energy, we are going to discuss here, has its influence in and out of our body. It acts in what we eat, we breathe, we take as medicines and we experience as environment.

Later, I could understand that the energy is used through a material to heal. Even though this system of medicine, I experienced, is not popular it is it is used by many people, thorough out the world. I realized then that these systems are simple and are a way to cure. This experience in medicine and its system made me explore more. It made me understand the difference between a material, energy, and energized material.

The negative ions, or the cosmic energy, around us engulfed me and urged me to read whatever I could lay my hands on that subject. The result is one of this book on cosmic energy. This book explains the energy around us, which is helpful in our health, food, and medicine. This book also explains what that energy around us is, in what form it exists, and what the sources are.

1. Introduction

Cosmic Energy is very subtle. This is the natural energy that is around us in everyday life. This can be utilized to improve our health in so many ways.

Energy is a common word. We cannot understand and go further unless we differentiate the energies in our lives. Food is the primary energy for humans, animals, birds and Sea creatures. Food is the material energy. There is another utterly non-material energy. They are light from the sun, visuals, music, etc., which help our minds. The third energy is one that connects the body and the mind, which we will talk about here.

We always give the importance to what we see and look at in front of us. We don't care what is hidden behind it. We don't give the importance to something hidden. Many things help us from behind. We talk about oxygen but don't care about the nitrogen that brings oxygen to our breathing. When a stranger brings a person involved an accident to the hospital, we care for the one who will be admitted. We don't give the importance to the stranger who helped this person.

We enjoy the fireworks only in darkness, but don't think of darkness. We cannot enjoy the Fire Works, if the place is fully bright. We can differentiate between letters and words only, when there is a space. But we don't think about the importance of space. In the same way, so many things help us directly or indirectly. But we don't realize them as we don't see them physically.

1. INTRODUCTION

Atmosphere air is the primary source of life. One such item in the atmosphere is the ionized air. As food is for our physical activity, the ionized air helps us with our mental activity. One of them, Negative ions, plays a vital role for our healthy life. Negative ion is cosmic energy that results from cosmic rays from the cosmos.

This book talks about the source of the cosmic energy and how our life is connected with it.

This book narrates some events in one's life as examples to show how cosmic energy plays a role in our daily lives. How life around us is hidden with so many facts we are unaware of it. Life goes smoothly with our experience of unknown facts unnoticed.

This book explains the atmosphere around us. It also describes how the winds create difficult situations for residents in various places on the earth. How can disturbance of the balance ions in the air cause so many ailments?

This book details the basics of positive and negative charges and how they are created. Also explains the source of both positive and negative ions—details how cosmic energy is produced and becomes part of the atmosphere.

The 'physics of cosmic energy' is presented in an Appendix for a technical reader. Those interested in technicalities can get detailed information about how ions are produced. Also, connected references would help.

This book briefs how cosmic energy can solve our health problems and how natural systems would emerge and fill the gaps in modern systems.

Human nature is to go fast in all activities. Man developed the material world to go faster in travel or communication, Nature has its own way based on its activities. The action speed of our body in breathing or digestion cannot be altered and not even in walking.

COSMIC ENERGY: WHAT EVERYBODY NEEDS TO KNOW

In the same way, the health and medicine systems have to go with the speed of our biological bodies.

Due to advances in science and technology we want to go faster than the speed of body in food and medicines. This creates lot of incompatibilities in our daily life. Understanding the hidden facts would help our lfe.

My presentation of points in this sequence may help any reader to understand nature and how to use it.

2. COSMOS

COSMOS- Universe:

When we gaze at the sky at night, not all the stars but many within the limitations of the human eyes are visible to us. Humans are attached to nature, and hence, everyone is attracted to the events of nature: the atmospheric conditions, the rain, the wind, lightning, etc. bring so many in us. The sky and the space beyond our view are the ones that create the eagerness for the human race to search more and more. The search goes beyond the orbit of the Earth, the Moon, and the Planets and even beyond the solar system. At the same time, we miss searching for subjects on earth that are useful for the human race.

The reason is to say that the energy in us comes from the sun and earth and so many heavenly bodies in the cosmos. The craving to see the space and enjoy it may be connected so.

The night is cool or warm based on where the individual resides. In warm countries or cold places, one tends to stay outdoors or inside the home but is eager to look at the sky. The sky is attractive with its stars and moon. The day sky with the shining sun is beautiful in cold countries and is not that very attractive in hot counties due to the glowing sun. The sky becomes attractive when it creates a rainbow, thunder, and lightning spectacle. The sky shows its colors through its clouds in the background of the sun and the horizon, etc. is gorgeous. We look outside when the ramblings from the inner voice are due to our problems, craving for solutions, and we need to feel free from the interior troubles. We also travel to places on earth to enjoy the products of nature.

Viewing the sun, moon, planets and stars have effect on us, in some way. Also the energy from them may affect us and also connected with us. It may link not only in the human body, the animals, and the roots of plants but in every cell and the elements on the earth. All elements in the universe are part of one close-knit family.

The moon revolves on its own and comes around the earth. In the same way, the earth rotates about itself and comes around the sun. The sun also rotates on its own and comes around the Milky Way Galaxy. The galaxy also rotates itself, and the process is endless in time and unlimited in space. When it comes down to the molecules, atoms and atomic particles also have rotations themselves and around others. How come everything rotates and is still stable? The reason is the existence of the magnetic field and gravity throughout the objects in space, from electrons to the stars and other large objects in the cosmos.

The stability is like a running bicycle movement. As long as its wheels rotate, the bicycle continues to run.

The stability of objects in space is connected to magnetic and electric fields. They are partners in establishing the links of bodies of cosmic space. So far, the limit of the cosmos and the objects in space has not been measured. Astronomical objects are stars, satellites, the Milky Way, Black holes, Meteorites and many more.

Cosmic rays are the source of from somewhere in the universe traveling to another place. Cosmic is a term often used in various spiritual and esoteric traditions to describe an unseen universal energy. It's believed to be the source of life for all subjects in the universe and the fundamental building blocks of reality.

Different belief systems have different interpretations of cosmic energy. It's thought to be a force connecting all living beings. It is

2. COSMOS

considered that the influence pervades all physical, emotional and spiritual well-being. Some believe harnessing the cosmic energy would lead to health and personal growth. Even though some assume that cosmic energy is a myth and there is no scientific evidence, cosmic energy has been proven to exist.

Objects in the cosmos are connected with space in between and separated by many forces, but they are also bonded one to the other. Even though cosmic objects are rotating themselves or cycling around others, they are independent. We can say that, they are connected statically with space in between.

Cosmic objects rotate themselves and move dynamically with some linked forces. Dynamically, objects in the cosmosphere are connected by cosmic rays, which are link between them. Every Cosmos object receives rays from somewhere, loses energy, and also passes further to many other objects. The energy flows by some rays, which we call it as cosmic rays. The energy we get out of cosmic rays is cosmic energy.

Cosmic energy travels everywhere, but the source is unknown. Maybe every cosmic object releases energy and goes to all other places across space, like helping each other. Mystery continues.

3. A Day in One's Life

Before we enter into the details of cosmic energy, let us look into experiences of individuals in their lives.

A person in hospital

> His father was admitted to the hospital in the Intensive Care Unit for about a week. The doctor said that his father did not need any more hospitalization and could move back to his house. His father was asthmatic and having breathing problems. The older man was on an oxygen support for some time. The doctor said his father would be fine if he started **breathing fresh air**.

Why use fresh air instead of pure oxygen for a patient? Why?

That is the experience with oxygen and air. One thinks fresh air is better than oxygen.

We can encounter many such events in our lives for which we cannot make any correlation. We see below some situations and events in one's life under various circumstances.

Temples, churches, houses, etc. which are built on hilltops are known to have some attraction for the people despite of the efforts to climb the steps. Why that attraction with hills?

3. A DAY IN ONE'S LIFE

Hilltop attracts tourists also. Why is that attraction, many build their houses and structures on the hill or mountains? Hills attract people even when the climate is the same as on the ground.

Children are eager to be on a terrace or climb the trees. Why?

Many passengers traveling by bus, car, or plane are known to feel uncomfortable. A journey by train never gives passengers that kind of stress. Why?

Train travel is more comfortable than bus or car travel. Is it comfort in a train, seating and movements, freedom in roaming on the train, or the scenery one enjoys?

It is known that some people tense up during lightning or thunder. The same feel subsequently relaxes during the rain!

What happens between the start of Thunder and lightning and then after rain.? Rain showers seem to promote relaxation of the mind.

It is narrated that dust and pollen cause Asthmatic problems.

Does the dust aggravate the symptoms of asthma or is the cause of asthma? Why does dust affect a few people selectively and not all?

Why food tastes better in some restaurants?

Even when food is prepared with the same ingredients, taste differs.

Beerhead (the frothy foam) or coffee foam tastes good. Why?

It is the same as beer, coffee, tea, or fruit juice. Why should it taste good if foam is present?

Walking barefoot gives a thrill. Why?

Starting with barefoot may create some difficulty. But as walking proceeds, there comes a feeling of walking more and more barefoot.

Above are a few examples. Many such instances can be narrated.

Many of us experience them but never feel there could be a link. Still, even if thinking arises, we find no time to analyze or no curiosity or capability to correlate. Many of the events above are experienced by you, me, and even our neighbor daily.

All the above events may not look necessarily related to one another. One may be surprised all the above-unconnected subjects have something in common to connect. If we go deep, we can see something in common. A feeling is created in those environments and atmospheres.

Some events experienced by our forefathers over a long time caused them to devise certain customs for their future generations to follow. Some customs remain in force and are still being followed, although many of us don't know why. Some customs and practices are modified by succeeding generations, and others are followed blindly without meaning or understanding. Some of the customs advised by our forefathers may be in good interest, learned from their experience and maybe with good outcomes. We do not follow them, as we don't reason out the meaning behind those customs. Even science might not have related the events that we and our forefathers are experiencing and have experienced.

Present science may not have the answers now, but the future may not be the same. The above events in nature that create the thoughts, feelings, and effects on humans are not in the realm of science. It is said that they have not been 'scientifically proven'. However, some scientists have started giving reasons for the above events and are helping others to utilize their benefits.

What is science? Is it Observation, Formation of a Hypothesis, Theory, or Rule? All observations on that specific category would agree with the formulated hypothesis. Due to scientist's research, the formulae and equations fit perfectly in physics and chemistry. By experience, the minute errors were cleared; perfections were

3. A DAY IN ONE'S LIFE

brought about by time. That is science, composed of measurements, readings, and records.

Our science is more oriented toward materials. That is why science doesn't delve into the area beyond materials. Humans are not materials, and many material theories don't fit well with human life.

Many such observations other than those of materials, such as the five senses of humans or animals and plants, may not be able to be measured. Hence, without those senses, a bio-system hypothesis is a complex process.

Human observation is built over experiences, shaping society's life. Slowly, society started living based on the experiences of others. Individuals started following society. Education and science need to observe and formulate theories. This resulted in meager inventions in the basics of life, nature, and the environment. Most of the new scientific research is over and above the existing concepts and repetitions.

The new thinking, even with new observations, has become very limited. The reason is thinking itself is channelized by education by teaching 'How to think'. Free thinking is reducing, and fundamental theories are in saturation, even though there is lot in nature that could be brought into science.

When a child is born and separated from the mother immediately after birth, the mother cannot identify her baby. But a sheep can identify its calf even in a crowd of lambs. Sheep have a sense of smell or something else. We may have lost many senses, but animals have not. Animals are with nature and nature makes them learn to survive.

When will science investigate, form hypotheses and theories for events like the Silva Method of mind training, one of the most powerful systems of self-help techniques? Out-of-body and near-death experiences are presented by Edgar Cayce, known as "The Sleeping Prophet," "America's Greatest Mystic," and palmistry by the most famous hand reader in history, Cheiro is yet to come under a theory. Water diviners could locate the water streams down below the earth. The phenomenon above may be called paranormal or parapsychology, but should be equated to the principle of science. Human observation and thinking have to go deeper and deeper to find the theories and solutions, and there may also be a limit for these Micro or Nano observations.

The human race has lost many senses through the course of channelized education and following. Science does not need to prove everything, and it can't prove it either. Science has limitations based on its types of equipment, parameters, measuring capabilities, etc.

Experience by animals and plants may extend beyond the limit of human observations, more in the areas where science cannot measure. Furthermore, it happens when humans cannot observe, but animals can sense. Beyond animals, there could be more and more signals that even plants can sense (Tompkins, 2002).

The police use dogs to sniff out drugs and trace the paths of robbers. Animals sense rain, earthquakes, and any natural events quickly. Scientists have yet to devise the theories and formulae to explain these.

People without access to technology from the development of science may still be in the Stone Age, but they live with nature and retain their old basic senses. Years back, during the tsunami in Indonesia in 2004, natives of the Andaman Islands in the Bay of Bengal Sea could escape to top of the mountains, and the non-natives, in the towns, perished.

Many scientists have proved some facts beyond the present science but cannot propagate their findings to the people. The reason is science psychology, which is called science addiction. Once any scientific information is embedded in our consciousness, the perception never changes quickly, and we feel that this is our final understanding and it is the right one to follow. Another idea in science is to enter into our consciousness and change the concept to the

3. A DAY IN ONE'S LIFE

core. The name Science does this tortuous task, and its success depends on the individual's acceptance or rejection.

Our human race is taught to follow the society. We believe in science when we are told or read 'this is scientifically proven' and 'science has established the facts.'

When an advertisement broadcasts that 98% of people get well from medicine XYZ, et cetera, we start consuming it. Why? What is the reason to believe in science unquestioningly? It is an addiction with knowledge. After a few years, the same science would tell us that the medicine XYZ produces a lot of side effects leading to bad health conditions. The government plans to ban the medicine only then. Even then, the same governments never cared about implementing these actions.

Think of Penicillin as an example. Penicillin is one of the first and also one of the most widely used antibiotics. This antibiotic was discovered by bacteriologist Alexander Fleming in 1928. This was the best antibiotic and saved many lives, mainly soldiers during the Second World War. Penicillin came to the medicine industry on a mass scale only in 1943 and started its side effects by 1947, within 20 years of its invention. Microbes started resisting the drug, producing toxins in the body and side effects like pneumonia. In 2005, the US FDA (Food and Drug Administration) banned penicillin in the poultry industry.

How do we differentiate different theories of science for the same subject with a gap of a few years? That is the experience of science. Any invention that is popularized gets instant attraction, and we believe it and follow it. We have no other way to survive.

The industry promoted the idea that synthetic fertilizers are the best and food production would reduce poverty in the world. A few years pass, the same scientists propagate not to use synthetic fertilizers and advise farmers to go for organic agriculture with natural manure. What was researched by a few scientists, years back, has to be proved wrong by the experiences of many.

Few years later, the concepts on the same subject may change. Now, the theory learned has to be unlearned. How do we differentiate those two theories? Which is wrong? The old or the new? Why one medicine that was good a few years back is treated badly now? We have to learn that what was good is no good now. Maybe after some years, this medicine may be considered reasonable. We accept the latest as correct, i.e., not learning but following.

Our life continued for centuries without any science. Life improved only by experience, and that experience was recorded by a few as observations and then theories came into in science. It helped us learn more and more and started binding us, making us forget our own experiences. Also, the same science has slowly started to watch the entry of anything new in science to be rejected or opposed if it hinders the existing business, cultural or individual interests. Even the proven experiments on biological transmutations by scientists like Kervran and others were rejected by other science groups (Jean-Paul 2012). Science has been taken over by business.

Science may not have the instruments and tools to measure many observations. The incapability to measure would ignore the observations as non-science or Para-science. There is tremendous scope for science to learn from nature and the area of research is wide open; when one can measure the very low frequency of thoughts and very high frequencies of the electromagnetic spectrum, when one can measure the very low light intensity near dark and high intensity of lightning, when one can measure very low currents of our body and high currents of lightning and when one can measure the distance between the particles in atom and distances between objects in space.

The limit for measurement has no limit, and one day, science may be able to record the observations if the science community helps society achieve that level of knowledge and intelligence.

When does the science would find theories with so many observations prevailing around us? How long and how deep could science be going into thinking, assessing, analyzing, judging, and deciding

3. A DAY IN ONE'S LIFE

on events in the life of day-to-day human activity? Now, life is beyond the grip of science for understanding. The grip of science is with mathematics and is a human invention. Next are physics and chemistry, which are connected with materials using calculations in mathematics.

With all the significant scientific developments, why engineering having theories held good for years? Why medical sciences are changing daily and even today, inventing new medicines for diseases? We do not know how long it would take science to bring biological sciences to its understanding.

Instead of bringing people closer to each other to live in love and harmony, practices in human society have started dividing them. Then again, reunite them in the name of Mother's Day, Father's Day and even remember and recall a lover on Valentine's Day. These special days are devised as a means to substitute genuine thoughts and emotions. The emotions have been lost in time from natural day-to-day activities in our social fabric. This is how we spoil first and then rectify in all spheres of human life.

Police and home security systems are there to install cameras. They will serve as deterrents or help capture the thieves. Years back, the same police advocated using a hard door or a strong gate so robbers would not enter. Many years back, people were cultured enough to be good in society, not to want other's wealth.

This is the advancement of technology and the degradation of human attitude.

Like fences for a land, rules are made for a society. More rules cause society not to think but to follow something unthinkingly. This is nothing but sheep followed by another sheep. As already mentioned, nowadays, people are coached how to think. It means the basic thinking process is on the path of decay with progress of AI, Artificial Intelligence.

We talk about health more after losing it. 'Prevention is better than cure'; it cannot earn money for the business industry if practiced widely. Hence, taking medicine is promoted as a better way to ensure good health than food. Eating processed sugar and taking a diabetic pill is the normal life of the day.

4. Cosmic Energy

In the previous chapter, we have seen some events a person experiences in daily life. What create an impact on all living things that are part of atmospheric air is the negative ions.

Cosmic rays travel across all celestial objects. When cosmic rays travel across the earth's atmosphere, their interaction with the molecules of air produces ions. One such ion is a negative ion, also called cosmic energy. The source of cosmic rays is not known. Cosmic rays are energetic particles and are mostly protons.

When any particle, an atom or molecule, loses an electron, the same particle gets a positive charge and is called a Positive Ion. The particle that receives the electron is negatively charged, called a Negative Ion.

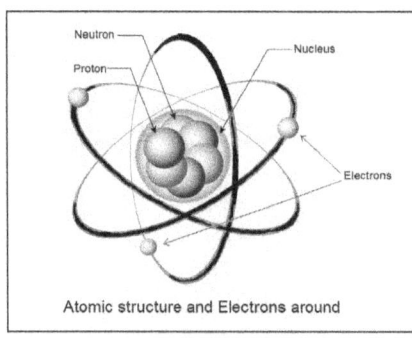

Atomic structure and Electrons around

Fig. 1

The smallest particle of an element, such as carbon, iron, silver, etc., is called an atom. A combination of atoms is a molecule. An atom (Fig-1) has three main components: neutrons, protons, and electrons. An Atom consists of a nucleus. The nucleus has neutrons and protons, whereas electrons rotate around the nucleus.

Neutron is neutral. Proton is positively charged, and electron is negatively charged. In a typical atom, the number of protons is equal to the number of electrons, and hence an atom is neutral.

Technical explanation is presented in Appendix-1, Physics of Cosmic Energy. It explains the events, production of ions, in detail.

We will see what these Cosmic rays are and how Negative and Positive ions are formed. This chapter briefly describes the course of events from cosmic rays to ions.

The cosmos is beyond the earth and solar system. Simply put, the particle that enters Earth's atmosphere from cosmic outer space is called cosmic rays.

> Assume a stone falls on a tree or somebody throws a stone to get a fruit from the tree. The tree is full of leaves and fruits. The stone falls on the tree from the top, hitting the leaves, fruits, and branches, and reaches the ground. More or less, all the energy would have been lost in its travel. On its way, the stone would have damaged the branches, made some leaves fall and some fruits to the ground. The stone will reach the ground with some energy or without much energy.
>
> If the stone doesn't touch any of the parts of the tree, the energy in the stone will be more or less the same. It will hit the ground with force and will lose all the energy in the ground.

When a cosmic ray enters the atmosphere of the earth, it loses the energy in hitting the particles and molecules in air that comes across in its way. Molecules in the air become damaged or change from their normal state. They take energy from the cosmic rays. Ions are formed, positive and negative.

4. COSMIC ENERGY

Ionized particles in the atmosphere serve many purposes, such as breathing oxygen for human beings, generating energy for plant growth, creating positive and negative clouds, forming thunder and lightning, etc.

Electrical and magnetic forces are the source of energy in human cells. The atmosphere gives us energy in the form of electrically charged ions. Out of both positive and negative, negative ion is an active energy for the life force. The old societies deeply analyzed human nature and had names for life forces. The life force in China is called 'chi (qi)', in India, it is called 'prana', Japanese call as 'ki', Greeks call it 'pneuma', and in Arabic, 'ruh' and 'Ruach' in Hebrew.

Apart from many Negative Ions due to cosmic rays, various sources exist for the charged ions and particles. Many manufactured industries such as cement, coal, thermal power stations, mines, etc., release gases with ionized particles as waste products. All types of gases released into the environment by industries and automobiles are waste products. Most of these from the sectors are positively charged particles. The bad Witches winds create more positively charged dust particles in the air. The dust and pollen from trees are other sources of positive ions.

Cosmic rays in the Earth's atmosphere:

The sun is considered to be a significant energy source for Earth. The solar system is believed to influence the Earth's atmosphere. However, Russian scientists Ermakov and Komozokov did measurements. They confirmed that ion production in the lower atmosphere due to cosmic rays is very high compared to that from sun. Hence, cosmic rays is considered as major player in atmospheric electric current, cloud formation, thunder, lightning production, etc. (Ermakov, 1992).

Cosmic rays, when enter into earth's space, it loses most of their energy in the higher layer of the atmosphere. The secondary rays enter the next level of atmospheric layers producing ultra-fine aerosols. These aerosol layers significantly affect the heat balance of the earth's atmosphere. The ionized clouds reflect the solar light upward and the earth's thermal radiation downward, creating a thermal balance. All recombination occurs in the ionization regions, because of the presence of free positive ions and negative ions in the same place. The recombination is very low at high altitudes due to high energy and speed. The recombination is high at the lower layer of the atmosphere because of the slow speed of ions and big-sized ion particles. Particle density is also high at low altitudes (Siingh, Devendraa- 2010).

Cosmic rays play an essential role in the ionization of the atmosphere than the sun's rays. Ionization of the region above the clouds has the impact of the charges ionizing the particles of the cloud. The interesting chain of reactions is that the ionized particles charge the clouds and clouds are the reasons for lightning. Lightning again produces charged particles of both types.

Nature creates ions in different ways. Below are the sources other than Cosmic rays in the production of ions.

- Sun and Ultra violet rays in sunlight.

- Earth's radioactivity.

- Natural radioactivity in rocks.

- The breaking of water droplets in waterfalls, sea waves, surfing, and rain.

- Electrical discharge during lightning.

- Friction due to air movement - winds and turbulence in the atmosphere.

Sun's energy:

We get light energy from the sun as a photon. A high-energy packet, a photon from the sun acts to knock out electrons from molecules in air

4. COSMIC ENERGY

and make them ionized. The ionization of air molecules occurs in the ionosphere mostly due to the sun's ultraviolet rays. The ionosphere is the upper atmosphere region as part of the thermosphere, 80 km (50 miles) from the earth's surface, extending to 700 km (440 miles). Sun's energy is intense in this region.

Earth's radioactivity:

A meagre amount of ions are formed from the earth's radioactivity. Earth has many minerals and natural elements, including isotopes. Most of these isotopes are radioactive. These radioactive elements ionize the atmosphere by adding more Negative Ions from their action on the earth's surface. Modern society by its concrete structures, hard pavements and metalled roads reduces the radiation flow and obstructs the path for releasing ions into atmospheric air.

Natural radioactivity in rocks:

Earth's radioactivity is in open space from the isotopes of mineral elements from mines and rocks. Radon, in gas form, is another source of ions that occurs naturally during the decay of thorium and uranium.

Waterfalls, sea waves, surfing, and rain:

We enjoy the waterfalls, water fountains, and bathroom showers. When air passes through water or vice versa, it interacts with water molecules, producing more Negative Ions than Positive Ions. This can be felt near the water waves hitting the seashore. The surfers enjoy the ride due to the comfortable sensation it creates. Light air flow in drizzling and during heavy rain causes a calm, soothing, and relaxing feeling. Waterfalls create freshness of vegetation. Just imagine the places people want to visit and roam around. Those are the places with more Negative Ions available that create a relaxed mind.

Lightning causes ionization:

Why lightning? The positively charged clouds get attracted by the earth. The reason, the earth is negatively charged. The clouds are of different categories, charged positively or negatively or without charge. When the positively charged clouds are close to the earth, there is an attraction between positive clouds and negative earth. In the same way, the joining of the clouds of opposite charges in the sky is the lightning above our head or far away on the horizon.

Current during Lightning:

100-watt light bulb draws around 1 amp (110V) or 0.5 amps (220V). Lightning causes a current flow of 5000 Amps just for a few seconds. That shows the strength of the cloud and how much charge is accumulated in the cloud.

The tallest building or a tall tree attracts the positive cloud, which makes a heavy current flow that makes an arc, which is lightning, and the sound out of it is thunder.

The lighting is the flash of arcing due to the high current discharge. This is like an electrical short circuit with a heavy current.

Air is ionized before lightning strikes, the reason for the high positive ions during lightning. Some people get a different feel during lightning. After the electrical discharge during lightning, most of the ions get neutralized. Some charged heavy particles remain in the atmosphere. Heavy particles are the ones, from which electrons are removed and positively charged. The rain brings down heavy particles to the earth. Small particles with a negative charge are unaffected. In this way, many particles with negative charges such as negative ions, remain in the air compared to positive ions.

Negative charges of the earth are peaked at sharp points of structures, tall buildings, and trees. That is the reason, tall buildings attract positively charged clouds.

4. COSMIC ENERGY

In the olden days, lightning demolished many tall buildings during the rainy season. Some buildings stayed stable even after many lightning strikes. Later it was found that the structures, uniform in construction and verticality, stayed without lightning strikes. A building with sharp tips on top of the structure helped the stability. Metallic tips were installed using this experience, and this design continued in churches, temples and mosques. The metallic tips and strips create a short-circuiting path with the ground, avoiding the current passing through the building and damaging it.

Winds and turbulences in the atmosphere:

When dry air moves along the land plains, it takes the dust, pollens and every positively charged particle along with it. When it hits the mountain surface, the dry air from the plains creates a rubbing action and creates ions.

The Negative Ions produced are less. Positive ions and positively charged dust consume most of the Negative Ions produced and are deposited along its way in land, houses, etc. Many such winds are explained in subsequent chapters.

The generation of negative ions in their concentration is closer to wind speed, water, plants, and air humidity. The most important factor is the water and the wind. The temperature of the air is less significant.

5. Cosmic rays- History

During many studies, scientists found energy rays. The rays from the Earth are considered to be through the earth and coming from the other side of the earth. Later, during many experiments, they found more rays from the space.

During research on energized rays, a German scientist, Victor Hess, raised a balloon at a height of 5000 meters and found that the radiation increased with increase in altitude. The radiation at high altitudes up was named as cosmic rays. His research on cosmic rays earned him a Nobel Prize. Victor Hess was awarded the 1936 Nobel Prize in Physics for discovering Cosmic rays in 1912. (Hess, Victor)

Earth is one of the cosmic objects we talk about here. Cosmic rays may dissipate all energy in the atmosphere itself. Some may spend their energy on earth. Some may cross the earth and travel further beyond Earth.

The magnetic field plays a significant part along with the negative and positive charges.

Magnetic force and electricity are active partners. A current flow through a metallic wire conductor can produce a magnetic field. Also, a conductor moving in a magnetic field can induce current. There is no interaction between magnetic and electric fields when they are static. When a metallic wire cuts a magnetic field, an electrical current flows in the wire. This is equivalent to a generator of electricity.

5. COSMIC RAYS- HISTORY

In the same way, when a wire with current flow, stays under a magnetic field, the wire would move akin to a motor. This is the active and dynamic effect of the two fundamental natural forces. Electric and magnetic fields are part of every object in the universe from the smallest atomic particle, electron, to the most prominent object in space. The magnetic field due to electricity or moving electrical charge is called the electromagnetic field. In static conditions, magnets have a magnetic field, and a voltage without current flow has an electric field (Robert O. Becker, 1998)

All the objects in this universe have their magnetic axis, around which they revolve. The living cells at the microscopic level also have a magnetic field with positive and negative polarizations.

History of observations and findings of ionic charges:

For centuries, people have known that, the air has various effects on people based on the climate, not only due to temperature and moisture but also something else. In the 5th century, Hippocrates observed in his place, the "Northern winds occasion disorder and sickness." (Guardian)

In 1748, French clergyman and physicist Jean Antoine Nollet invented one of the first electrometers, the electroscope, which detected the presence of electric charge by using electrostatic attraction and repulsion. He had experimented by planting mustard seeds in two pots. Using an electrostatic generator, he passed the positive charges to one pot and nothing to the second pot. After about a week, all the seeds planted in the first pot sprouted and showed their growth by a few millimeters, whereas the second pot showed very little progress compared to the first pot. (Rossi, Bruno)

In the late 18th century, Giuseppe Toaldo, a priest, was appointed the chair of astronomy at the University of Padua. Toaldo paid great attention to the study of atmospheric electricity. He became a famous physicist and professor. He observed the plant growth has some link with the electricity that exists in the air as static electricity. He observed the plants growing next to a lightning rod grew almost ten times taller than identical plants just a few meters away. Based on the views of Benjamin Franklin of the US he promoted the erection of lightning rods as a preventive and protective action. Through his pressure, lightning conductors were installed to protect the tall structures of Siena Cathedral, on the tower of St. Mark's in Venice. Venetian navy also adopted his method of installing lightning arresters on ships. (Giuseppe Toaldo)

In 1780, Abbe Bertholon of France observed some electrical states in the air caused changes in people's health. He noticed that vegetables grew an extraordinary size when water was electrified by an electrostatic generator and poured on them. Also, he invented a unit to collect atmospheric electricity using an antenna and pass the ionized air onto plants named "Electro-vegeto-meter". (Electro-vegeto-meter)

Even though many people have proved static electricity has its say over plant growth and is also documented and demonstrated, understanding the reasons was not easy, no one was able to explain how it happens.

In 1899, German physicists Julius Elster and Hans Friedrich Geitel proved that electrostatic fields were based upon the existence of electrically charged particles called ions. To prove this, experiments were conducted by a team at the Air Ion Laboratory of the University of California.

By passing electrically charged particles over the plants, it was proved that the growth rates of the plants are high, and the ions have a physiological effect on the plants.

Dr. Clarence W. Hansell of RCA Laboratories was the first to find out the ions could also impact individual's state of mind. His observation occurred in 1932, when he noticed a change of mood in a wild way of a colleague working near an electrostatic generator. This incident caused Dr. Hansell to monitor his colleague continuously. He found on another day that his colleague was cheerful and happy when the generator was made to produce negative ions. Subsequent tests proved depression and irritability, when the machine was producing positive ions. Dr.Hansell's observations widened, and started noticing the effects of atmospheric ions on individuals. (Clarence_Hansell)

Serotonin and positive ions:

During the late 1950s, research by Dr. Albert Krueger proved a small number of negative ions could kill all types of bacteria that caused colds, influenza, and respiratory infections. Then he tried with a large group of mice with various concentrations of ions. During his experiments, he found overproduction of Serotonin when the mice were subjected to positive ions, leading to hyperactivity, exhaustion, anxiety, and depression. The same group with negative ions was introduced after stopping the positive ions, which created calmness among them with serotonin reduced. This research uncovered the mechanism fundamental to altering moods by altering the types of ionized air. He found reliable data that there was a significant and consistent reduction in blood levels of serotonin when the mice were exposed to air with Negative Ions of 400,000 per cc.

The hypothesis, he concluded was the high dose of positive ions increases the serotonin levels, and the high negative ions stimulate the action of speeding up the metabolic removal of serotonin.

Serotonin is a vital chemical molecule produced in the body and is popularly thought to contribute to feelings of well-being and happiness. It also plays a significant role in plants and animals. Serotonin plays a significant role in the human central nervous system, such as depression, mood, perception, anger, aggression, behavior, anxiety disorder, social phobia, memory, sexual behavior, appetite, and sleep. In addition, serotonin has essential functions outside the central nervous system, including regulating energy balance and food intake and activities of gastrointestinal, endocrine function, cardiovascular, and pulmonary physiology. Serotonin is connected with several human diseases like depression, schizophrenia, migraine, hypertension, and eating disorders.

It was found that the serotonin in the brain is a factor representing positive ion levels. Negative ions help to increase mental activity by diminishing serotonin levels. Patients who are affected by positive ions go into a state of depression and bad moods. They are treated with medicines of the type of serotonin blockers. The medical field has contradictions. The reason for increased serotonin levels is not apparent whether serotonin causes mental depression or mental depression causes an increase in serotonin. The subject of serotonin and its effects are still in debate (Blenau, 2015).

6. Life Around us

The physical objects around us influence our lives rather than other non-material subjects. Many non-material subjects exist, such as vision, music, speech, lectures, natural sceneries, etc. They are just through energy medium without any physical contact. Hence, we don't give importance to their effect on us.

Food is the primary source for the body energy. Oxygen plays a significant role in food consumption. When a cell uses food energy, oxygen helps to burn the carbon in the food. The combination of oxygen and carbon is the source of energy for the cells.

Physical things connected with us have more attractions than what is hidden around us. We don't see the hidden, we don't feel their presence and hence we do not understand much about them. Also, we do not understand much about our lives; they are controlled by unknown subjects rather than material subjects. We think more about our body than what controls or helps our body.

We do not understand that non-material energy helps us. Energy by way of hearing, vision, reading, etc. is used without much importance given by us.

We think more about the food we eat than the air we breathe. In the same way, we talk about air and oxygen rather than some hidden subjects in the air that help our life.

The hidden one in the air, we breathe, plays an important role in our life, mainly in our thinking and activities of mind.

Negative ions

We read about ions in the previous chapter. We will see its importance in human life.

Comfort for you and your body:

When sitting in a room, you feel uncomfortable even when you are not sweating and with the right temperature and humidity. If the room size is small, you are more likely to feel uneasiness and discomfort. The feeling arises less often, if the room is large and with a high roof. You may think the room is air-conditioned flawlessly, and adequate air is available, but why should you still feel stuffy?

What happens in reality?

Even when the same temperature and humidity prevail in different rooms with the same cleanliness, you may not feel the same level of comfort in all the rooms. Sometimes, we say fresh air circulation is missing.

If the room is clean and has fresh air circulated with the air-conditioning system at an optimal temperature and humidity, no uneasy feeling is supposed to come up. At the same time, when you stand outside the room and in the open atmosphere with the same climatic conditions, your feeling is incredible. Why?

The Simple reason is the feeling of well-being depends on outside air, which has something to do with energizing you to make you comfortable in feeling and physical presence. Room air conditioners and HVAC systems are designed to allow a small quantity of outside air. Some HVAC (Heating, Ventilation, and Air-Conditioning) systems are not designed to add sufficient outside air to the re-circulated ventilation system.

6. LIFE AROUND US

Why do you need outside air circulation when the surrounding air has the correct temperature with the required airflow around your body, and the air is humid within comfortable limits? Think for a while. There is something that we are not aware of, but it makes us feel comfortable when the outside air is mixed with the re-circulated air. The re-circulated air had some energized particles at the start, which were consumed by the people in their rooms. The charged particles in the air create comfort and clear our uneasy feelings.

The experience is more dramatic when a crowd gathers in an auditorium. There is freshness when few people arrive and gather in the hall. In a big hall or theatre, the atmosphere is good and people would start gathering. The event begins in time, the program continues, and slowly, people feel suffocated. At first, nobody feels the itch. The clock ticks and everybody starts consuming the energy in the atmosphere from the air present, and the refilling of fresh air is limited. Our assumption is oxygen consumption. We are unaware of the reduction of negative ions apart from oxygen.

We think oxygen is consumed, and carbon dioxide is released. This does not happen, where there is less crowd or a good amount of circulation of outside air mixed with conditioned air (Wallach, 2010). That is why HVAC has to be designed to provide good fresh air for the ventilation of closed areas.

The charged particles in the atmospheric air create the comfort that the temperature and humidity cannot produce. The charged particles that make comfort are called Negative Ions. The atmospheric air also contains Positive Ions. The imbalance between the charges creates discomfort. The balance is when positive ions and negative ions are within specific proportions. The imbalance is caused when positive ions are increased in ratio to the Negative Ions.

Ions in one cc :

One cc (cubic centimeter) of air contains 2.7×10^{19} molecules.

1250 positively charged particles with 1000 negatively charged particles in the air are in normal condition. 2250 ions out of this quantity is nothing compared to the total number of molecules. Best ratio of Positive Ions to Negative Ions is 5:4. This ratio of around 5:4 is the best atmosphere for an average individual.

Harmful environments increase the Positive Ions to a higher ratio even upto 12:4.

The extra entity in the air, negative ions, promotes good feeling. We start consuming the Negative Ions in the air and need refilling. The re-circulated air in the room never brings the same quality as fresh air, which contains the Negative Ions. Better than the re-circulated air in the room, a step into the outside environment brings more comfort because of the availability of Negative Ions.

All the suffocation is due to insufficient Negative Ions in the closed room.

Let us briefly look at what happened in the events "A Day in One's Life" seen in a previous chapter.

Fresh air and the air from the open atmosphere will improve the hospital's health, not the pure and clean oxygen.

Oxygen only gives energy to the cell. The dynamic activity is due to the negatively charged particles. i. e . Negative ions in the air help mental activity. Hence, a patient feels more comfortable with fresh air than oxygen and atmospheric air has something more than oxygen.

Temples, churches, and even houses built on hilltops are known to have some attraction for the people despite the efforts to climb the steps. Is the attraction more towards structures on the hill or mountain than those at the ground level?

6. LIFE AROUND US

Hilltops have a high Negative Ion ratio, so there is an attraction to reach the hilltops. With that experience, our forefathers built holy places and residences over the hilltops.

Many passengers feel uncomfortable when traveling by bus, car, or plane. A journey by train never gives passengers that kind of stress.

The Positive Ion collections become high on the moving vehicles due to static electric charges. This is due to the high interaction between the vehicle and the air. The positive charges collected get discharged due to the metallic wheel of the train. The cars and buses with rubber tires have no path to discharge. The accumulated charge creates a high Positive Ion ratio.

It is known that some people tense up during lightning or thunder but feel subsequently relaxed during the rain!

Approaching rain with lightning and thunder: Hair-raising feelings happen during lightning and thunder due to the presence of highly positive ions.

Also, when clouds of positive ions are close to the earth and near us, elated feelings will arise, such as euphoria. By the time drizzling starts, the tense feeling dies down, and brings a chilly feeling from the Negative Ions.

Some people get affected due to high Positive Ions, which create a tense feeling during lightning. Lightning produces a lot of ions, which is also more positive. The high ratio of Positive Ions during lightning is the reason for that tense effect. When the same amount is reduced during rain, relaxation happens.

It is known that dust and pollen cause or accentuate asthmatic symptoms.

Asthmatic people are mentally sensitive and hence, need more negative ions in the air, they breathe than others. They breathe more air where the atmosphere lacks negative ions. Breathing more than normal air causes an obstruction by way of some growth in the nostrils and air passages.

They feel comfortable with the fresh air of the countryside. The dust and pollens are mainly positively charged. This dust and pollen attract and mix with negatively charged particles in the air. This action brings down the needed Negative Ion ratio.

Asthmatic people suffer at midnight around 2 AM. This period is the time atmosphere has lowest negative ion ratio. Best ratio is around 30 minutes before sunrise.

Why food tastes better in some restaurants?

Apart from ingredients, the method of preparation makes a change. When food ingredients are ground, oxidation happens, and taste comes in.

Beerhead (the frothy foam) or coffee foam tastes good. Why?

It is the same beer as coffee, tea, or fruit juice, but the foam is filled with ionic charges and increases when rinsed.

Walking barefoot gives elation. Why?

Starting with barefoot may create some difficulty. But as walking proceeds, there comes a feeling to walk more barefoot.

Morning and Evening

When a person gets up early in the morning, there is freshness at sunrise, and the same person, when he gets up after a nap in the evening, feels dull. One can notice this in the morning, when their house is built with its front facing east, causing them to be cheerful. Many temples are built facing the east. Many prayers are performed by looking at the morning sun.

6. LIFE AROUND US

Only some locations would be more comfortable sleeping in the same room in a house. One can find a better negative ion ratio in that area if analyzed.

The good and bad of a place are learned by gaining experience over a long period.

7. Wind around us

A person's experience with nature is not his age or years he dwelt in that place or locality. One has to add the total years the person's ancestors and his generation learned over the years in that geographical area with the surrounding atmosphere, vegetation, people, natural events, etc.

The same atmosphere that gives energy also creates health problems. There are many dry winds or 'Witches' winds in different parts of the world. In nature, at many places on earth, the air, though it may not be contaminated or polluted, develops an imbalance due to the electrical charges of the particles in the air. The dry wind may be hot or cold, but the ionic particles charges in the air change. It happens mainly in places where mountains and air rifts with the sides of the mountain and the valley channel the airflow.

These winds are typically warm and dry. The air currents occur where mountains stand in the path of solid airflow. The rubbing of air along the slopes of the hills creates electrical charges of the particles in air. This is like rubbing the palms of one's hand, which makes hair raising due to the build-up of static charges.

One such significant wind is from the European Alps, known as the Foehn, a poison wind. Similar winds occur in many parts of the world. Many of such winds are named as it plays a significant role in human life. The standard wind that does not affect us much goes unnoticed without anybody's curiosity. The winds get a name

7. WIND AROUND US

when they are famous or notorious for causing discomfort, havoc, disasters, and problems to the community.

More attention is given to any factor when it causes an issue that profoundly affects us.

Many winds with different names exist worldwide, such as the Bise, Foehn, and other Witches winds.

Bise wind in Geneva, Switzerland is a cold, vigorous, and persistent north or north-easterly wind blowing from the alpine mountains. This affects Switzerland and eastern France. The city of Geneva sits on a narrow V-shaped plain between Lake Geneva and the point where the Alps converge. The Bise is especially prominent in the Geneva area at the southwestern end of Lake Geneva. Fred Soyka describes his health experience in Geneva due to the dry atmosphere in his book 'Ion Effect'. (Soyka, Fred)The Bise whips up the surface waters of Lake Geneva and in winter, the sea spray encases the nearby vegetation under thick ice. During the summer, the dry Bise is a perfect laundry drier, finding its reflection in the local folklore: "Avec la bise, lave ta chemise", when the Bise blows, wash your shirt. During the dry season, when winds create an uneasy atmosphere and discomfort, people move to the rural areas. Doctors are busy too this season due to the bad health conditions of the inhabitants.

Foehn in Geneva, Switzerland is different from Bise winds. Foehn is dehydrated warm and dry wind is from the southeast, touching the surface of the mountains and hitting the valley, the lake, and the city. Foehn is a condition rather than the state. The state is felt after the positively charged particles are left in crevices and packets of the streets and homes.

These charges create a build-up of static electricity and stay long even after the wind dies down. In the 19th century, Austrian physician Anton Czermak published a clinical review of the effects of Foehn as the residents in areas of frequent Foehn winds were reporting illnesses ranging from migraines to psychosis. Also, a study by the University of Munich (Ludwig-Maximilians-Universität-München) found that suicide and accidents increased by 10 percent during Foehn winds in Central Europe.

Foehn in Munich, Germany and Austria is a generic term for warm, strong, and often arid downslope winds that descend in the lee side of a mountain barrier. *Foehn* (föhn in German) type winds are known for their rapid temperature rise, desiccating effect and the rapid disappearance of snow cover.

Tramontane in France is a strong, dry, cold wind from the north or northwest. In the Mediterranean, the tramontane is created by the difference in pressure between the chilly air of a high-pressure system over northwest Europe and low pressure over the Gulf of Lion in the Mediterranean. The high-pressure air flows south, gathering speed as it moves. The continuous howling noise of the *tramontane* is said to have a disturbing effect on the human psyche. Victor Hugo, in his poem, the main character says, "The wind coming over the mountain will drive me mad..."

The Mistral in France is an intense, cold, and north-westerly wind that blows from southern France into the Gulf of Lion in the northern Mediterranean. Though it blows in all seasons, the mistral (meaning 'mud eater') occurs mainly in winter or spring.

It lasts only one or two days and sometimes continues for more than a week. It has a significant influence all along the Mediterranean coast of France and frequently causes sudden storms in the Mediterranean. The mistral is usually accompanied by clear and fresh weather, which plays a vital role in creating the region's climate and bringing good health. The dry air dries stagnant water and mud and blows away the pollution from the skies over the large cities and industrial areas. The mistral also saves crops from the spring frost. Also, the sunshine and dryness carried by the mistral wind have an essential effect on the local vegetation. The vegetation already due to rainfall shortage is made even drier. Because of this, vegetation is more prone to fire, and once the fire starts, the wind escalates the fire. The mistral causes irritating headaches, and mothers complain

7. WIND AROUND US

that the wind provokes restlessness in children. Some pet keepers complain that pets are affected by this wind.

Bora in the Adriatic is northern to the north-eastern wind in Montenegro, Italy, Bulgaria, Greece, Slovenia, Bosnia, Croatia, Herzegovina, Bulgaria, Greece, Poland, and the southwest of Russia. In Croatian, the wind is called "burno", which means "violently" and is commonly used to describe the weather. Bora is a cold and arid wind. These winds occur any time of the year but mainly in the cold season, lasting from a few days to a month.

Halny wind in Poland blows in south Poland and Slovakia in the Carpathian Mountains. A wind comes from the South on one side, down the slopes of the Tatra Mountains in Slovakia, and from the North on the other side of the mountains. Most Halny occurs in October and November, sometimes in February and March, and rarely in different months. Halny is a warm wind through the valleys and is often disastrous.

It causes avalanches, lifts the roofs of homes, and, according to some people, can influence the mental state of some people, like the Foehn wind.

Sirocco in North Africa is the name for the hot and humid wind in some parts of North Africa, and it has many names in each region. The hot air originates directly from the Sahara desert along the northern African coast, producing hot, dry, and dusty conditions. The desert air moves over Northern Africa and flows northward into the southern Mediterranean basin. Visibility becomes very poor, and the fine dust that blows might result in damage permeating into the instruments and pieces of equipment. On rare occasions, the Sirocco picks up enough dust and sand to produce even sandstorms. Sirocco creates cool, wet weather in Europe. The wind is mainly in March and November and touches a speed of 100 Km/h (160 mile per hour) at its peak. Sirocco creates a depressive feeling among people during the wind season.

The Sirocco has different characteristics and has many other local names, too. The term Sirocco is not used in North Africa, where it is called **chom** (hot) or **arifi** (thirsty); **Simoom** in Palestine, Jordan, Syria, and the desert of Arabia; **Ghibli** or **leveche** (or Chibli, Gibla, Gibleh) in Libya; **Chili** (or Chichili) in Tunisia and Algeria; **Sharav** in Israel; **Ikslok** in Mediterranean; **Khamsin** (or Chamsin, Khamasseen) in Egypt and around the Red Sea.

In one of its translations, the Old Testament calls the Middle Eastern Sharav an evil, destructive and deceiving wind.

Sirocco in Europe affects the southern Mediterranean basin. It has the same demoralized feeling that prevails in the North African region. Lousy health conditions in individuals create trouble between people.

Simoon in the Sahara and Arabian Desert is a hot, dry, suffocating wind combined with dust that swings across the African deserts, mainly Arabia, Syria, Jordan, and its neighboring countries. *Simoon* creates extreme heat in the prevailing arid deserts and the sandy plains. *Simoon* lasts less than half an hour and brings up the dust and sand from the floor of the desert. *Simoon* is not the primary wind. This is a second wind that results in thermal heating of the dry surface. This wind reshapes the terrain and forms dunes, too. The name comes from the Arabic '*samma*' for Poison. It is also called *Samiel* in some regions.

Khamsin (fifty) in North Africa is a Fifty-day wind is an oppressive, hot, dry, and dusty south or south-east wind occurring in North Africa, around the East Mediterranean and the Arabian Peninsula intermittently in late winter and early summer, but most frequently between April and June.

A counterpart of the sirocco, it is a southerly wind over Egypt blowing from the Sahara Desert and an easterly over the Negev Desert and parts of Saudi Arabia. The term is also applied to powerful southerly or south-westerly winds over the Red Sea. Less frequently, the *khamsin* might also occur in winter as a cold, dusty wind.

7. WIND AROUND US

Santa Ana winds in California and the seasonal strong wind in Southern California is hot, dry, and dusty. The Santa Ana winds originate from inland, sweeping down from the deserts and across coastal southern California, pushing dust and smoke of forest fires far out to the Pacific Ocean. Santa Ana winds blow mostly in autumn and winter but can arise at other times.

Depending on the prevailing temperatures in the source regions, they can range from hot to cold. The winds are known mainly for the hot, dry weather, often the hottest of the year and are infamous for fanning regional wildfires. For these reasons, they are sometimes known as the "devil winds" across Southern California. It is widely believed in Southern California, that the winds negatively affect people's moods and behavior.

Even without ironclad scientific proof, it is a well-accepted part of local lore. Some believe the winds also create positive ions, which are believed to affect mood negatively. Many of them consider this to be the cause for the statistical increase in the number of suicides and homicides during these times. The Santa Ana wind also becomes a part of novels like the "Red Wind" by Raymond Chandler and Ross MacDonald. American Indians, according to their mythology, call the Santa Ana wind Bitter Winds. The folklore tells of people's wind sickness, when exposed to the Bitter Winds.

Chinooks in the US blow across most of the North American states, particularly the Rocky Mountain region, during winter. Statistical studies point out that during these winds, road accidents happen more frequently and suicide rates increase. Hospitals postpone some operations and wait for the winds to calm down.

Chinook in Canada, interior West of North America, refers to the winds where Canadian Prairies and Great Plains meet various mountain ranges. The name Chinook for the wind is derived from the name of the people of that region, along the lower Columbia River. A vital Chinook can make snow one foot deep almost vanish in one day.

Chinook winds have been observed to raise winter temperature, often from below -20 °C (-4 °F) to as high as 10-20 °C (50-68 °F). This lasts for few hours or days, and then the temperatures plunged to their base levels. Chinook winds, sometimes, are said to cause a sharp increase in the number of migraine headaches suffered by the locals and are often called "chinook headaches". At least one study conducted by the Department of Clinical Neurosciences at the University of Calgary supports that belief. They are popularly believed to increase irritability and sleeplessness. Many migraine sufferers believe weather changes can trigger migraines.

Thar Desert, Rajasthan-India, has a hot and dry environment due to the isolation of the desert by the mountain ranges and plains. This contributes significantly to the weather patterns that shape its distinctive environment around the desert. The desert effectively absorbs all the moisture carried in the monsoon clouds before the clouds can reach the desert. The resulting monsoon winds in the desert are hot and dry. Wind activity over the Thar Desert is at low ebb during the winters. From March onwards, when the surface is dry and temperatures soar, the summer winds associated with southwest monsoon reach a maximum speed of 20 km per hour or more. Advancing summers further dry up the ground, and whatever vegetation tries to grow in the winters is dry by May. The arrival of the monsoon in July puts an end to this activity.

Aadi Kaatru (*Aadi* wind) in South India is popular due to the dry atmosphere, which creates dullness among people. The wind force is high and used to be a shrill noise. This is the season for the Southwest monsoon. If the rain joins the wind, the damage would be high.

The monsoon has two branches of wind, the Arabian Sea Branch and the Bay of Bengal Branch, near the southernmost end of the Indian Peninsula. The wind hails from the Bay of Bengal and travels on the eastern side of the mountains of Western Ghats.

This is primarily a dry wind. The Arabian Sea branch carries the sea's moisture and blows towards the North, creating an excellent

7. WIND AROUND US

rainy atmosphere in North India that goes up to the foothills of the Himalayas. The wind from the Arabian Sea travels west of the mountains in Kerala, but drags the wind of the Bay of Bengal. This wind dragged rubs along the mountain bordering Tamil Nadu. The wind blows in the northern direction, traveling a long stretch of the hills. The air carries heavy dust and sometimes forces people to stay indoors. The wind is forceful, during July and August and sometimes with rains. Due to the dry wind, a lousy mood prevails in most parts of the Southern Indian Peninsula.

After monsoon rain and plowing, the farmers in the southern part of India do the sowing in mid-August. This sowing time is based on experience gained over time. Another question is why sowing is so precise at this time (Srikumar-2014).Hence, it is considered an inauspicious month, too. No major decisions on any event would be discussed in the family.

Family functions, carnivals, temple functions, business deals, and marriages will be avoided during this period. Even decisions or discussions on events like marriages are avoided. A newly married couple would not be allowed to live together (Sankaran, 2014).

Like Bise, Foehn, Tramontane, Bora, Sirocco, Simoon, Khamsin, Santa Ana winds, Chinook, Thar desert wind, and Aadi wind these winds in different parts of the globe have the same, but similar characteristics of the Foehn wind of Europe.

These winds have other names in many other regions: Zonda in Argentina, Koembang in Java, Warm Braw in Schouten Islands off the north coast of New Guinea, and Oroshi in Japan.

Common in all the winds is increased positive ions compared to negative ions.

8. Positive and Negative ions

Positive and Negative ions:

Life without problems may not be attractive. Like a coin, every subject and event in the world has two sides, and even thinking in life has another side, and some events may have many sides. The two-sided events coexist and cannot be separated. An example is a symbol called Taijitu (see Fig) for Yin Yang. Yang represents the bright day, male, etc., while Yin represents the dark, night, female, etc. One is mingled with the other. Happiness in life depends on how we utilize both sides of everything. It is not about how it is. It is the way, we use them.

While we use the sun and its brightness and spend more time working during the day, we rest for a few hours of the night in darkness. We optimize and utilize both. We don't reject the other side and cannot reject it even if we do not like it. If one rejects it, nothing exists for that person, and life is imbalanced. There cannot be a coin with just one side.

We have to live with both the positive and negative aspects of everything but with balance. Living with an atmosphere of both positive and negative ions in balance is life. The words Positive

8. POSITIVE AND NEGATIVE IONS

and Negative have nothing to do with positive benefits and negative effects on the person. They are just arbitrarily identified as positively charged and negatively charged ions and one opposite to other.

Negative ions have more beneficial effects on human health than positive ions. Negative ions make our minds calm, tranquil, and relaxed. One enjoys the traveling or driving on the roads of ups and downs and at hillside. One will enjoy turns, slopes, rises, and falls. How boring it will be driving on a plain road. At the same time, the ups and downs shall be within limits for enjoying the journey, say within certain ratio.

Like travel, life shall have ups and downs to enjoy a rhythm, like a piece of music, winning and losing, running and resting, sleeping and waking, etc.

There shall be an excellent matching balance of positive and negative sides. That is where one enjoys life. Positive and negative are necessary, but within limits and in better comparison ratio.

Ion Comfort with Positive versus Negative Ion:

As already explained in a previous chapter, the comfort of a human being is, when both positive and negative ions are present in a particular proportion. 5:4 is the best ratio of positive and negative ions. For 5 Positive Ions, 4 Negative Ions are normal and comfortable. Every individual has different requirements for ions. Because of these reasons, not all people are affected by the increased positive ion levels.

Some unventilated areas increase the Positive Ions to 4 times than the Negative Ions, such as in closed rooms, cars with doors closed, room air-conditioners with locally circulated air, industrial areas with dust, during approaching storms, etc.

Out of both ions, Negative Ions, what we call the Cosmic Energy, enable us to maintain a balance with the positive ions and play an essential role in our health.

COSMIC ENERGY: WHAT EVERYBODY NEEDS TO KNOW

Full moon and positive ions:

The moon reflects the sunlight to its maximum during the full moon. The moon's surface has Negative Ions like the earth's surface. The moon reflects the photons from the sun without much angle deviation and hence, energetic. This negative charge of the moon by the photonic effect towards Earth repels the outer layer of Earth's ionosphere, a field of Negative Ions. In this way, the ionosphere is pushed close to the earth's surface. This creates a higher positive ion population than the existing negative ions. The full moon effect is due to the increased positive ions.

Agriculture - Seeding and planting:

Investigators have found that plants are sensitive to negative and positive ions. As per Old Farmer's Almanac, planting trees and many crops around the full moon makes them grow better. This early growth also gives better yields later. Vegetables like lettuce, tomatoes, carrots, etc., planted during the full moon time, grow faster and healthier than those planted on other days. Scientists reason that there are more positive ions during this time of the full moon, and these ions play an important role in water absorption by plants from the roots. Positive ions play a role in the subsoil level for the sprouting and inner core part of the plant, whereas negative ions help in the outer layer of the plant.

Seeding on a particular day in a specific month by the farmers in that locality is based on their experience—the time of seeding yields better produce. The choice in the time and the day of the month is based on their experience with or without understanding the presence of positive and negative ions. An abundance of positive ions during the Seeding is essential.

Positive ion helps the seed to germinate. Once the leaves see the sky, negative ions play a significant role in plant growth. Farmers in Southern India sow the seeds during July,15-Aug,15. The wind at that time creates an abundance of positive ions. Seeding is done

8. POSITIVE AND NEGATIVE IONS

at that time and thus, positive ion helps the seed to germinate. Subsequently, the monsoon rain takes care of the plants.

With all this negativity, sowing done during that period makes better germination due to positive ions and sprouts protruding the soil smoothly. Positive ions have an essential effect on the germination of seeds, and negative ions help plant growth. After this period of germination, the atmosphere changes and is full of negative ions and help to grow the plant.

9. Wind and Health

Health effects of the winds - Good and Bad:

You may wonder how different types of winds in various weather conditions affect our health, mainly our mental health. The warm, dry wind is likely to drain the energy of the people, making them suffer daily. *Witches* Winds have been blamed for making the people bad-tempered, irritable, prickly, depressed, tired, sleepy, and unhappy. The same has also been blamed for increased quarrels, traffic accidents, murders, and suicides. Many countries suffer due to the winds of their landscape. All the troubles are mind-related and reflected in the body too. The winds themselves are problems due to the damage of properties, etc., and the effect on people's mental depression is due to the content of the wind and the high value of positive ions compared to the negative ions. The imbalance of ions, the molecules of air within the winds, makes people so frustrated and annoyed.

These invisible ions are inhaled and also adsorbed through the skin. They are either positively or negatively charged. When the air has too many positive ions, which are 1,800 times heavier than the negative ones, it becomes dry and heavy. The heavy positive ions were carried away by raindrops and cleaned away. At the same time, during the rain plenty of negative ions are generated. The Negative ion is increased due to the interaction of raindrops with air. Negative Ions would increase to 3 times that of Positive Ions after the storm and during rain (Wallah, 2010).

9. WIND AND HEALTH

Many hill and mountain areas, seashores, and waterfalls produce large amounts of Negative Ions. Yosemite Valley, California, has one of the most Negative Ion-filled environments in the state. In New York and Canada, Niagara Falls has been called the most immense Negative Ion generator globally. In our homes, a great Negative Ion generator is the bathroom shower.

What does the dry atmosphere create in the health of the people of that region? The Swiss Meteorological Institute extensively studied in 1974 and published the problems arising from *Foehn*-type wind. Above all, these winds can make us seem bad and short-tempered, argumentative, and quarrelsome without any apparent reason. Also, this makes us think that we are in good mood even though it is not so. Table 1 lists the problems in various places.

When the dry winds die, they leave another effect due to the static electricity. Static electricity is formed due to the accumulated concentration of charged particles in one place. Positive ions are heavy particles and are also attracted to the negative earth. The accumulated charges create an electric field similar to the state of a high-voltage system creating an electric field. These electric fields create disturbances to biological systems.

A build-up of static electricity occurs due to the accumulation of negative or positive ions. Individuals can feel static electricity when they touch the metallic door-knob after a walk over synthetic carpets. Bad ventilation in computer rooms also creates static charges. Hence, floor carpeting is done for them with anti-static properties, carpets ingrained with metallic fibers and connected to the floor.

Winds cause	Subsequent effects
Anxiety and tension	Leading cause for bad decisions
Catching cold	No patience to wait and creates interest in medication.

Lassitude, sleeplessness, or bad sleep	Lethargy
Depression	Lethargy
Reduced sex drive	Broken marriages
Changing moods	Non-cohesive decisions
One day optimistic next day depressed	Alternating moods
Exhaustion	Higher incidence of heart attacks
Respiratory problems	Asthma
Irritability	Slower reaction time
Body pains	Joint pains
Sick headaches	Nausea, stuffy nose
Variations in body salts - sodium, calcium and magnesium	Dizziness

Table-1. What Positive Ions create and subsequent effects during witches' winds.

Until then, scientists believed that the earth's radiation ionized the air. During and after Albert P. Krueger, a Bio-meteorologist and Professor Emeritus at the University of California, Russian scientist A.L. Tchijewsky started 1933 serious experiments. They reported that the growth of some bacteria is inhibited when air has a high content of negative ions.

To substantiate Dr.Krueger's findings, Dr. Felix G. Sulman, Head, Department of Applied Pharmacology at Hebrew University in Jerusalem, discovered the primary cause behind the unpleasant symptoms experienced by people in the desert region during the period of poison winds. He could link the people's moods during the

9. WIND AND HEALTH

winds of Sharav to the winds of Sirrocco in Italy, *Foehn* in Central Europe, and Santa Ana in California (Soyka, 1991).

Health during a *Foehn* wind:

In a 1980 study, Dr. Felix Gad Sulman collected urine samples from 1000 volunteers regularly just before the arrival of the storm of *Foehn* wind and normal weather conditions.

Compared to normal weather conditions, the samples taken just before the *Foehn* wind showed increased production of serotonin. He concluded that, the high concentration of positive ions carried by these winds would stimulate an increase in the production of serotonin and histamine in their bodies, causing allergies, migraines, difficulty breathing, irritability, and anxiety.

Also, the samples were analyzed for hormone levels, adrenaline, etc. The production of adrenaline initially induces a state of euphoria and hyperactivity. Hyperthyroidism was also associated with increased positive ion conditions. Although serotonin is exceptionally crucial to the functioning of our bodies, he concluded that these individuals, who are weather-sensitive, produce too much serotonin during stormy winds. They were affected by their serotonin. Thus he coined the term as "Serotonin Irritation Syndrome."

Dr. Werner Becker, a Professor, Department of Clinical Neurosciences at the University of Calgary in Alberta, studied for two years, keeping a record of the health conditions of many patients during pre-Chinook, Chinook, and non-Chinook days. Out of 75 patients, 32 were likely to have migraines during the Chinook. Dr. Coleman and Becker said the weather was main migraine trigger. Chinook is a *Witches* wind in which positive ions are higher.

To overcome all the harmful effects of the atmosphere and *Witches* winds, nature provides us with abundant Negative Ions.

10. Cosmic Energy in Health

When an individual is on the top of a building, the person feels elated. There is also an interest in walking up the stairs or going by elevators to be at top of a building.

Why?

Charged accumulation at peaks

Children want to climb trees, even with all the risks. The interest is to climb up and up. Reaching the tree top is more exciting, and the enthusiasm, the tree top creates, is achieved by imparting energy to the climber. There is enjoyment too, when sitting under the branches of the tree. Tree houses are another way of inviting the tree to feed energy to the sleepers.

There is something very subtle that creates an interest in people to go up the hills. When people climb hills and mountains, the feeling of tiredness is less and nullified by the interest in climbing. The reason for this entire inner urge to climb is the presence of increasing negative ions, cosmic energy, up the hill.

The earth is the biggest negatively charged mass. Touching the ground with bare feet and hands makes one feel contented. That is why most passengers get off a bus during a brief halt en route and walk on the ground. The land surface imparts a sense of relief from the stress induced by travel. In the same way, kids are interested

10. COSMIC ENERGY IN HEALTH

in playing in the mud, which gives them a lot of energy. It is not the clay with which they wish to play; it is the soil of the earth with which they like to play, no matter how messy it is.

The atmosphere, by way of air, gives us oxygen to get the energy and power for the living cells to keep the bio-system (body) alive. But something else is needed to keep the mind energetic and to act. That is the Negative Ions, part of the air, which are working their magic on us without our self, knowing or realizing what it is.

Modern Living:

Our modern life made our homes, offices, public places, and transport systems modern but away from nature. This made the environment unhealthy. Modern buildings have enclosed rooms, ventilation systems with forced air and artificial electrical lighting emitting electromagnetic radiation. Synthetic industrial products replace natural materials and contemporary office gadgets like computers and mobiles with a wide range of radiating energies and waves.

Simple Methods to utilize the cosmic energy in Modern Living:

With all the comforts of modern life, the attraction of nature never leaves us alone.

- Air conditioning systems (Charles Wallach) shall be designed to allow external air into the ventilation system so that fresh air brings Negative ions.

- Make fresh room to enter closed rooms once in a while.

- Enjoy the waterfall, water fountain, rain showers.

- Walk over the soil with our feet touching the earth.

- Move and Walk around the environment of trees. Breeze through the trees.

COSMIC ENERGY: WHAT EVERYBODY NEEDS TO KNOW

How to use cosmic energy in Food:

Food is the fuel supply to the body to energize all the cells of the body. Food can provide energy for the body. But food alone cannot make up for brain activity.

The way food is prepared is important than the contents in the food. Method of preparing food is essential to input the negative ions into food.

- Cooking in open vessels.

- Avoid pressure cookers.

- Cook slow with reduced heating.

- Grind the spicy ingredients instead of simply using them. Grinding adds charges. Brings taste.

- Grinding herbal leafs than just adding in food.

- Fresh water has ionized oxygen. Boiled water loses oxygen. Boiled, cooled water can be rinsed to improve oxygenation.

- Rinse the water as much as possible.

11. Cosmic Energy in Medicine

The medicines used in modern medicine systems are manufactured materials and primarily synthetic chemicals. These medicines are in the material category. Modern medicine systems use the material directly.

We talk here about medicines made from natural materials. The source for making medicines may be herbs, food items, or any other material. Medicine from these materials is made to fit human nature.

We talk here about medicines of energized nature.

When a material is imparted with energy, its action is not the same as the original material. For example, think of some tools we use, such as crowbar, needle, and arrow. Can these tools do the same job as iron rods, wire, and wooden sticks or more than that? Do they act same or differently?

An iron rod becomes a crowbar. A wire becomes a needle. A wooden stick becomes an arrow to work with nature. Even though many consider humans as physical bodies, Humans are a product of nature, a subject with some sort of life similar to plants and animals.

When a herb or medicine material is treated with negative ions, it takes the ions and takes a different form. The ionized product becomes a tool to carry the ionization and acts as an energy or charge and not as a material.

Negative ion generators help the asthmatic patients. They create an ionized atmosphere even though they cannot match the natural sources. The generation can be balanced with positive and negative ions. A negative ion generator does not act as medicine but as a palliation. These negative ion generators helped scientists, doctors, and other researchers to experiment with people to learn more about the basics of ions and their use.

Appendix-1
(Physics of Cosmic Energy)

Cosmic and Sources of ions:

We get light energy from the sun as a photon. Other than this, we get energy through Cosmic rays from space or say, the cosmos. Cosmic rays travel across all celestial objects. The source of them is not known. Cosmic Rays are energetic particles and mostly protons.

During many studies, scientists found energy rays. These were thought to be from the Earth. Later, during many experiments, they found more rays were from space. The rays from the Earth are considered to be through the Earth and coming from the other side of the Earth.

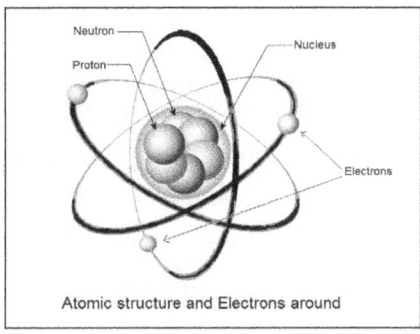

Atomic structure and Electrons around

Fig. 1

Cosmic space is bombarded with cosmic rays, which create ions of a Negative and Positive nature.

APPENDIX-1 (PHYSICS OF COSMIC ENERGY)

In this Appendix, we will see how.

Elements

Before we understand cosmic energy, we should know some basic elements of our lives. There are 92 basic elements, from Hydrogen to Uranium. They exist in the subjects of earth. Cosmic Ray interaction is at the level of an atom of an element. Hence, we have to understand the contents of the atom.

Elements are pure substances with specific properties. Atoms are the most minor units of elements that still retain the element's properties. Atoms contain electrons, neutrons, and protons. Each of an element is defined by the number of protons in its nucleus.

Atom-Electron-Proton and Neutron:

An atom (Fig-1) has a core and its surroundings. The core is called a nucleus, which contains neutrons and protons. The electrons are circling the nucleus. Proton is positively charged, and electron is negatively charged. Neutron is charge-free. Every atom has an equal number of protons and electrons, which is electrically neutral. The number of neutrons in most atoms usually is equal to or higher in number than the protons. The exception is hydrogen, which has only one proton with an electron and no neutron.

Hydrogen has one proton and one electron. Oxygen has eight protons, eight neutrons in the nucleus, and eight electrons surrounding it. Carbon has six protons, six neutrons in the nucleus, and six electrons surrounding it. An element is defined only by the number of protons called the atomic number.

The Neutral State of an Atom:

A typical atom remains neutral, when the number of protons and electrons is the same. The value of the charge of an electron is the same as that of a proton except that it is negative (or opposite i.e., positive). When an atom in a molecule or compound of molecules

COSMIC ENERGY: WHAT EVERYBODY NEEDS TO KNOW

combines with an atom of another molecule, sharing the electrons, a strong bonding is created, and the combination becomes another molecule.

The resulting molecule has no free electrons. The charge gets neutralized in an atom with equal protons and electrons.

The combined weight of protons, neutrons, and electrons decides the weight of an atom and the weight of that metal or gas. Both hydrogen and oxygen are gases that combine to form a liquid, called water. This way in nature, we have atoms and molecules, a combination of atoms. Two hydrogen atoms and one oxygen atom form water, and the weight of water is 1 gm/cc.

Charge of an Electron:

> One Negative ion has a charge of 1 EV (Electron Volt) and equals $1.60217657 \times 10^{-19}$ coulombs. EV is a basic unit to measure small charges. One Coulomb equals 6.3×10^{18} EV.
>
> 1 Coulomb=63,000,000,000,000,000,000 EV
>
> 3600 Coulomb has a capacity of 1000 mAH (Milli Ampere Hour) AA or AAA size 1.2 V chargeable cell, which has a capacity of 500 to 3000 mAH. You can imagine how small an EV is.

The number of protons in the nucleus decides the type of element, either oxygen or carbon or any other. The proton would remain the same in a component irrespective of neutrons and electrons.

All the subjects on earth consist of one or more of these elements and their combinations of these 92 elements. These elements are fundamental to the living system on Earth.

The atmosphere consists of oxygen, nitrogen, carbon, etc. Our human body consists of many elements: carbon, hydrogen, oxygen, iron, magnesium, calcium etc.

APPENDIX-1 (PHYSICS OF COSMIC ENERGY)

Cosmic Particles:

Any particle from the space entering into Earth's atmosphere must interact with atmospheric air. Cosmic rays emanate in space from the objects in the universe and enter every object in space. One of the celestial objects, earth, utilizes the particles from space and by a natural process. Cosmic rays have different actions in different layers of atmospheric air. Cosmic rays are mainly of high-energy protons of hydrogen nuclei, alpha particles (two protons and two neutrons like helium atoms but without electrons) few heavier atoms, and a smaller amount of electrons.

The energy of electron and a cosmic particle:

> The energy of one electron is called electron Volt (eV). The energy of a cosmic array is in the range of 50 MeV and 20 GeV (20 Giga eV -20,000,000,000 eV).
>
> Compare an electron's energy of 1 eV to that of the energy of a cosmic ray.

In travel (Fig.2) cosmic energy loses its energy in every atom on its way. The loss is based on the type of atom and the way, it hits or interacts. The electrons removed from an atom may be one or more. Overall, the atmosphere is ionized by the interaction of cosmic rays. The energy of cosmic rays is reduced when they reach the earth's surface. Finally, it may reach the earth with some energy or lose all energy in the atmosphere.

In the course of cosmic rays crossing the atmosphere, electrons from the elements are knocked down. The element deficient of electrons becomes positively ionized. The one which receives the electron becomes negatively ionized.

For example, oxygen becomes positively ionized when an electron leaves the oxygen atom. When an electron joins a nitrogen atom, it makes nitrogen negatively charged.

COSMIC ENERGY: WHAT EVERYBODY NEEDS TO KNOW

POSITIVE and NEGATIVE ions are created in the atmosphere. Both positive and negative ions exist in the atmosphere in certain proportions.

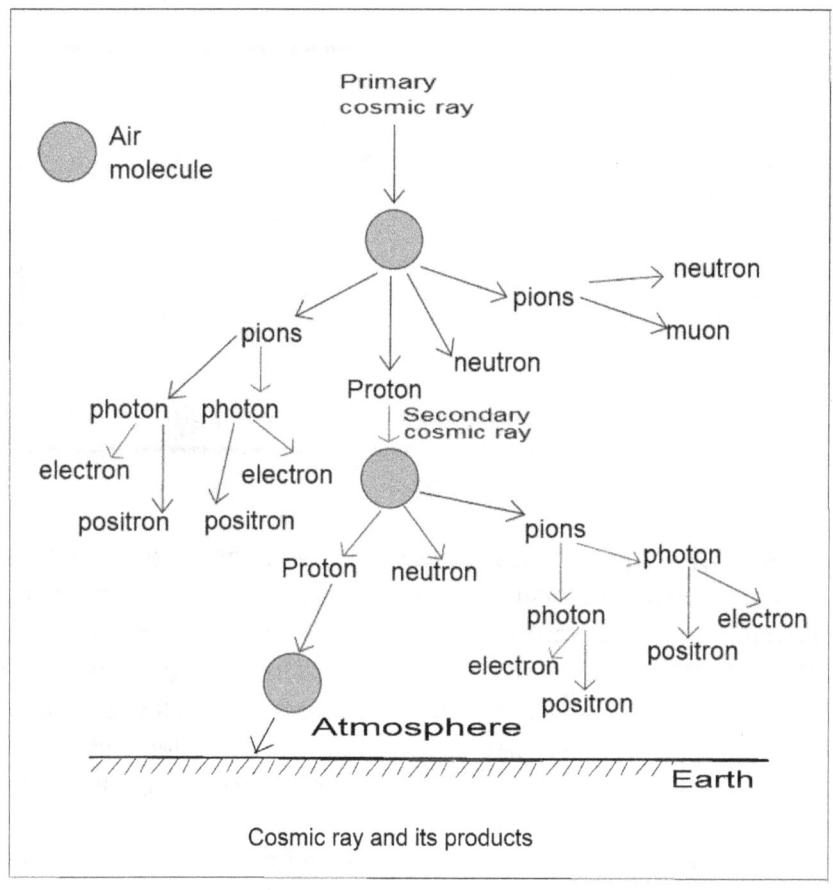

Cosmic ray and its products

Fig. 2

APPENDIX-1 (PHYSICS OF COSMIC ENERGY)

What exactly happens, when cosmic particles collide in the thermosphere and further? The particles hit the nitrogen, oxygen, and carbon dioxide molecules in the air, and electrons are knocked out (Fig. 3).

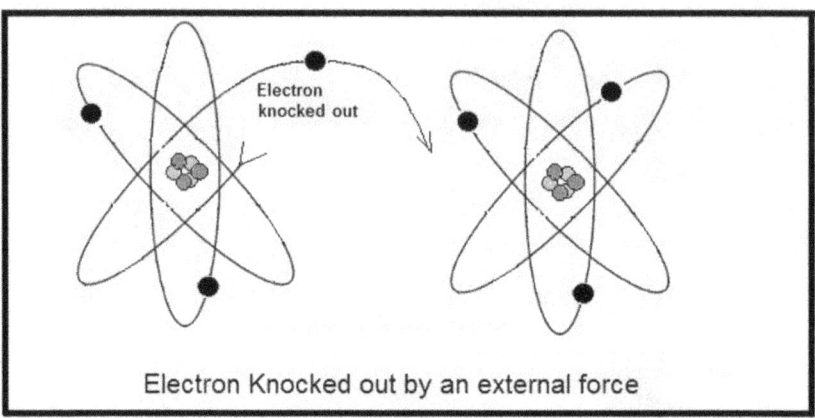

Electron Knocked out by an external force

Fig. 3

The stability in the electrical charge of an atom or molecule means the atom never gets disturbed. A molecule or atom maycombine with another atom or molecule to form a new molecule. Yet, no ionic charge is created. The atoms or molecules share electrons between them. When an electron or electron in an atom is attracted to another atom and combined, a molecule is formed. This is called a chemical process; the elements stay together, and the new molecule is stable.

When an electron is knocked out from an atom, the electron is free. The knocked-out electron is free and joins with another atom. The atom that receives or acquires an electron becomes negatively charged (see Fig.4). The atom that gets the electron becomes negatively charged.

The charge is equivalent to the charge of an electron. By this, the loser or the electron donor gets positively charged. These ionic particles are due to the shortage or lack of electrons or extra electrons.

These Ions may be atoms or molecules or particles with a combination of so many molecules. Negative Ions are small molecules that are free to wander in the air. It may get attractedto positively charged particles, and the particles will become neutralized.

We have ionized particles with positive and negative charges in the atmosphere. The atmosphere consists of air that contains oxygen, nitrogen, carbon dioxide, and many other gases. Any of these elements may become ionized.

The negative ions that are part of air become part of our breathing and part of every biological life on earth and are called cosmic energy.

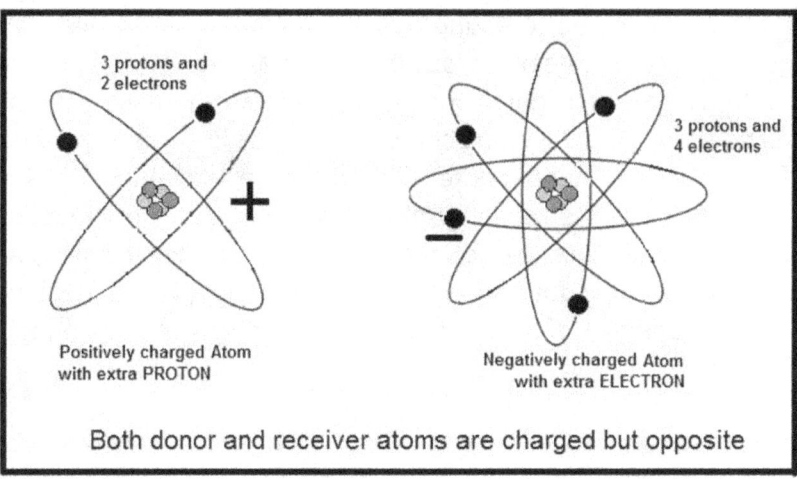

Fig. 4

Cosmic ray travel

High-energy packets, known as photons, travel from the sun at the speed of light. Similarly, Cosmic rays travel through space at a very high speed, the speed of light. Similarly, all the cosmic particles collide first at the atmospheric layer, the thermosphere.

The energy of the primary cosmic particle is so high, it creates a shower of charged particles, when it collides with components of air. The cosmic particles travel further with reduced energy and are called secondary cosmic particles. Some cosmic particles reach

APPENDIX-1 (PHYSICS OF COSMIC ENERGY)

earth without much hindrance but with reduced speed after many collisions.

More air ionization occurs in the Stratosphere and Troposphere, up to 50 km (30 miles) from the earth's surface. The electron knocked out from molecules of air may be one or more. The molecules that lose electrons are now positively charged. The free electrons in the air combine with the other neutral molecules of nitrogen, oxygen, and carbon dioxide and form negatively charged particles. Whenever more than one electron is knocked out from one molecule, one Positive Ion is created, and more electrons are released. They join with other molecules, creating more Negative Ions. In this way, Negative Ions are more numerous than Positive Ions. Most electrons join with oxygen, the lighter of all active elements in the air. At the same time, the Negative Ions combine with positively charged dust and make a clean atmosphere. One of the beneficial functions of cosmic energy is the cleaning of atmospheric air. The ionization has a lot more effect on global temperature and the climate, air precipitation, cloud formation, creation of aerosols, ozone, and electrical properties of the atmosphere, including lightning. (Siingh, Devendraa- 2010).

The atmosphere protects life on Earth. The atmosphere absorbs most of the infrared and ultraviolet rays and curtails the effect of high-energy cosmic rays. The high-energy particles, called primary cosmic ray, will collide with the oxygen or nitrogen atoms in the atmosphere and then become secondary particles.

These secondary particles have sufficient energy to produce the next generation of particles. Successive transformations produce a large cluster of particles from both ionic charges. (Fig-2) However, most of these particles get absorbed or decayed in the atmosphere. Very few cosmic particles reach the Earth's surface. The energies of the particles hit the surface of the Earth are very low when it approach the Earth's surface. Through these acts, the harmful effects of high-energy cosmic particles are reduced. This is the way the atmosphere acts like an umbrella to protect bioactivity.

What is the source of cosmic particles in space? Baade and Zwicky suggested in 1934 that cosmic rays originate from supernovae. In 1948 H.W. Babcock suggested that the magnetic variable stars could be a source of cosmic rays. Y. Sekido and his colleagues suggested in 1951 that cosmic rays emanate from the Crab Nebula. Research is still in progress, with acceptance and contradictions from scientist's suggestions. Suggestions of the sources for cosmic rays include supernovae, active galactic nuclei, quasars, and gamma-ray bursts (Walter, 2012)

The most significant source for ionization is the cosmic particles, which are a little from the sun. The earth's atmosphere's creation of ions is much less than that of the cosmos.

Electronic charge in electricity for an understanding.

In electricity, we get power from the movement of electrons.

As long asthe charge remains in an atom, it is static. Dynamic movement or the flow of charges or electrons causes electricity. When electricity is stored in batteries, it is in the form of charge. There is a current flow when we connect a load to the battery outside. This is like storing water in a tank, and using the water through a tap is an electric current.

Our lives depend greatly on the atmosphere, so we have to care about it. The atmosphere gets graded in high and low in various parameters when we go up and up. The atmosphere extends only up to a certain height over the earth's surface; beyond that, the air is absent.There are different layers (Table-1) of the atmosphere, categorized based on their state and conditions of temperature, temperature gradient, air density, and air contents.

Troposphere	0 to 12 km	0 to 7 miles
Stratosphere (also ozone layer)	12 to 50 km	7 to 31 miles

APPENDIX-1 (PHYSICS OF COSMIC ENERGY)

Mesosphere	50 to 80 km	31 to 50 miles
Thermosphere (includes Ionosphere)	80 to 700 km	50 to 440 miles
Exosphere	700 to 10,000 km	440 to 6,200 miles

Table-1 Atmospheric layers

The layer troposphere extends from the earth's surface to 12 km (0 to 7 miles). The distance of the troposphere from the Earth varies from the poles to the equator. The layer extends up to 8 Kilometres at the poles.

The troposphere is the layer for life on Earth, and the air is denser, containing 75% of the mass of whole atmosphere. The world's weather takes place only in this region.

The temperature decreases to a level of -60°C at 12 km. Temperature increases in the Stratosphere by increasing altitude and reaches from 60°C to 0°C. The rise in temperature is due to the absorption of UV rays from the sun. This layer has no air turbulence like the troposphere and is cloud-free.

The mesosphere is the third layer. By increase of the altitude, the temperature decreases to -100°C. This is the coldest of the layers of the atmosphere. This is the layer, where the meteors burn when they enter the atmosphere, and shooting stars are seen.

The thermosphere is the hot region with low air density. The sun heats the atmosphere, raising the temperature to 1000°C. Due to the sun's high impact and low air density, ionization occurs in this region.

What do we have in this atmosphere to survive and keep the plants and animals energetic apart from just surviving?

Air contains primarily oxygen as a fuel for energy production for life on earth.

Major constituents of dry air are (Fig.5) Nitrogen, Oxygen, Argon, Neon, Helium, Krypton, and Xenon; except for Nitrogen, CO2, and Oxygen, the others are inert gases. Inert means that, there is no chemical reaction with other elements. Apart from the above gases, atmospheric air is polluted with many gases from chemical and other manufactured industries. Oxygen and Carbon-dioxide are primary in feeding energy to the living.

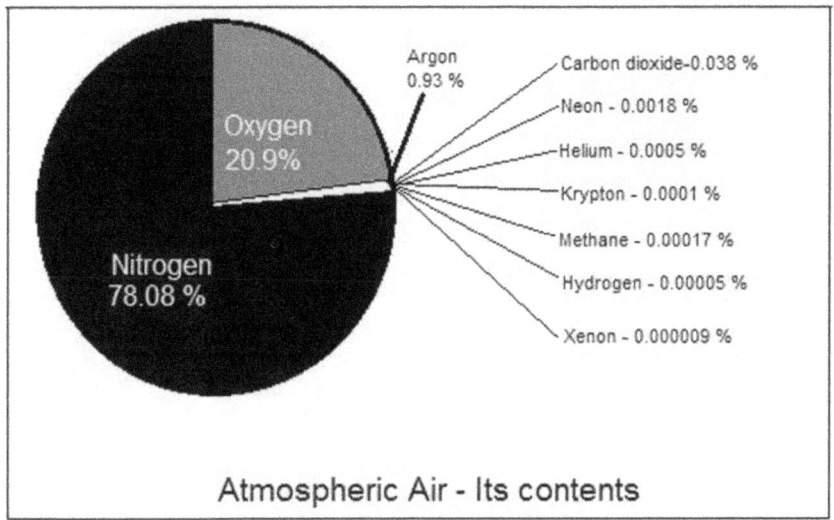

Fig. 5

The atmosphere feeds us with cosmic energy as Negative Ions. With these charges in space, our life is intermingled and influenced by the balanced nature of positive and negative ions.

References

- "American Health demographics and spending of Health care consumers" –New Strategist publications-2010

- Becker, Robert O. – Body electric

- Bise winds of Geneva - www.weatheronline.co.uk/reports/wind/The-Bise.htm

- Blenau, Wolfgang - Seratonin receptor technologies - Humana press Springer-2015

- Bora - www.weatheronline.co.uk/reports/wind/The-Bora.htm

- Dr. Charles Wallach – "Ion Controversy – A scientific Appraisal" -2010

- Chinook and health – www.migraines.org/about_media/helthsct.htm

- Chopra, Deepak – "Perfect health" -1990

- Clarence_Hansell - https://en.wikipedia.org/wiki/Clarence_Hansell

- Cosmic rays -en.wikipedia.org/wiki/Cosmic_ray

REFERENCES

- Duke, James A. - Dukes's handbook of Medicinal Plants of the Bible –CRC press -2008

- Electro-vegeto-meter
  https://www.jstor.org/stable/43426439

- Ermakov,V.I et al - Ion balance equation in the atmosphere – Journal of Geophysical Research – Vol 102 Oct. 1997

- Foehn - www.summitpost.org/foehn-effect/466432

- Full moon effect - www.policeops.com/full-moon-ion-effect.htm

- Guardian
  https://www.theguardian.com/uk-news/2017/nov/21/weatherwatch-the-winds-that-brought-good-or-ill-health

- Giuseppe Toaldo https://en.wikipedia.org/wiki/Giuseppe_Toaldo

- Haboob - www.weatheronline.co.uk/reports/wind/Haboob.htm

- Hayanon- 'What are cosmic rays' translated by Y.Noda et al.

- Hess, Victor- https://en.wikipedia.org/wiki/Victor_Francis_Hess

- Ions - www.hyperstealth.com/ions.htm

- Ionisers –

- Jean-Paul – Journal of Condensed Matter Nuclear science 7 -2012

- Ken Wilbur- "The Spectrum of Consciousness"

- Mistral- www.frenchentree.com/holidays-in-france/weather/the-infamous-mistral-wind-in-provence/

- Natasha - www.newscientist.com/article/dn3228-air-ionizers-wipe-out-hospital-infections.htm

- Natasha McDowell – "Air ionisers wipe out hospital infections" –New scientist -3 January 2003 by

- Negative Ion Report: The CBS Nightly News, Feb 14, 1995

- Negative Ion generators - www.negativeiongenerators.com/negativeions.html

- Norman cousins "Anatomy of Illness" by Rane Dubos in preface.

- Dr. Kruger A.P.-Report on negative ions-1967

- Robert M. Schoch, Robert Aquinas - Pyramid Quest: Secrets of the Great Pyramid and the Dawn of Civilization

- Rossi, Bruno- "Cosmic Rays: A Dramatic and Authoritative Account"

- Sankaran- www.indiastudychannel.com/forum/128075-Karkitakam-adreaded-month-for-Kerala-people.aspx

- Seratonin - www.causeof.org/sis.htm

- Siingh, Devendra and Singh RP-"The role of cosmic rays in the Earth's atmosphericprocesses" - Pramana-Journal of Physics – Jan 2010

- Srikumar – Kolar Gold field, (unfolding the untold)- 2014

- Soyka, Fred & Alan Edmonds (1991). "The Ion effect – How air electricity rules your life and Health", Bantam Books

- Thar desert www.newworldencyclopedia.org/entry/Thar_Desert

- Tompkins, peter - "The Secret life of the plants" -Earthpulse press-2002

REFERENCES

- Ullman, Dana – "The Homeopathic Revolution: Why Famous People and Cultural Heroes Choose Homeopathy" - North Atlantic Books – 2007

- UMM -

- Universal Energy- www.cosmicloti.com/tag/universal-energy/

- Wahlin, Lars – Atmospheric electrostatics, Colutron Research coprortion-1985

- Walter,M –Early history of cosmic particle physics - European physics journal June-2012

- Welton, Janus Aia Bbei - "The Living Elements of Healthy Building Design

- Winds effects - www.nytimes.com/1981/10/06/science/ions-created-by-winds-may-prompt-changes-in-emotional-states.html

- Zapping Airborne Salmonella and Dust - published in the March 2000 issue of Agricultural Research magazine.

www.ingramcontent.com/pod-product-compliance
Lightning Source LLC
Chambersburg PA
CBHW020515030426
42337CB00011B/399